UNDER THE CI

By the same author:

Collected Poems 1987 (OUP, 1987)
Selected Poems 1990 (OUP, 1990)

UNDER THE CIRCUMSTANCES

Poems and Proses

D. J. Enright

Oxford   New York
OXFORD UNIVERSITY PRESS
1991

Oxford University Press, Walton Street, Oxford OX2 6DP
Oxford New York Toronto
Delhi Bombay Calcutta Madras Karachi
Petaling Jaya Singapore Hong Kong Tokyo
Nairobi Dar es Salaam Cape Town
Melbourne Auckland
and associated companies in
Berlin Ibadan

Oxford is a trade mark of Oxford University Press

© D. J. Enright 1991

First published in Oxford Poets
as an Oxford University Press paperback 1991

All rights reserved. No part of this publication may be reproduced,
stored in a retrieval system, or transmitted, in any form or by any means,
electronic, mechanical, photocopying, recording, or otherwise, without
the prior permission of Oxford University Press

This book is sold subject to the condition that it shall not, by way
of trade or otherwise, be lent, re-sold, hired out or otherwise circulated
without the publisher's prior consent in any form of binding or cover
other than that in which it is published and without a similar condition
including this condition being imposed on the subsequent purchaser

British Library Cataloguing in Publication Data
Enright, D. J. (Dennis Joseph) 1920–
Under the circumstances : poems and proses. – (Oxford poets).
I. Title
821.914
ISBN 0-19-282834-7

Library of Congress Cataloging in Publication Data
Enright, D. J. (Dennis Joseph), 1920–
Under the circumstances : poems and proses / D. J. Enright.
p. cm. — (Oxford poets)
I. Title. II. Series.
PR6009.N6U48 1991 821'.914—dc20   90-46359
ISBN 0-19-282834-7

Typeset by Wyvern Typesetting Ltd.
Printed in Great Britain by
J. W. Arrowsmith Ltd, Bristol

Acknowledgements

Acknowledgements are due to the editors of the following periodicals and anthologies, in which some of these poems and proses first appeared: *Acumen, The Beloit Poetry Journal, Encounter, First and Always* (Poems for Great Ormond Street Children's Hospital), *The Listener, London Magazine, London Review of Books, Mother & Child* (Birthright), *The Orange Dove of Fiji* (Poems for the World Wide Fund for Nature), *PN Review, Poetry Book Society Anthology 1989–90* and *1990–91, Rhinoceros, The Spectator, Thames Poetry, The Times Literary Supplement*.

Contents

1

Demonstration 11
In the basement 11
Aria 12
Aye 12
Out and about 12
Carrier bag 13
Money talks 13
City sorrows 14
Assimilation 14
One can't go on feeling guilty for ever 15
Prose 15
Tatami 16
Starboard home 16
Going Abroad?: a leaflet 17
From little acorns 17
As you say 19
Computereality 20
In High Dudgeon 20
Door 21

2

And 23
Eve's first confusion? 23
In a corner 24
Secret history 24
Paradise retained 26
Kingdom come 27
Homo felix 28
Having the last word 28
And again 29
Disciple 30

3

National anthem 31
Bonefires 31
Here, there 32
Besieged 32
Museum of Atheism 33

Act of faith   34
The guarded tourist   34
Every bullet has its bill   35
Cambodian legend   35
Matter of form   36
Peace and war   37
Seminar on contemporary Chinese writing   37
National Character: Variations on a theme of Roy
   Fisher's   39
'Why does almost everything seem to me like its own
   parody?'   40

4
Announcement   41
Babies   42
Case record   43
Dolls   43
Hell rejuvenated   44
Second childhood   44
Poem on the Underground   45
Wonders of Nature   46

5
'You have to commit a poem the way you commit a
   crime'   47
Market forces   47
Takeover   48
Blurbing   49
Library book   49
Another anecdote from William IV Street   50
Primitives   51
Event   51
Bad behaviour of language   52
Recipe   52
Logistics   52
Letterheads   53
Pet   53
One too likes books   54
Qwerty etc.   54
Dear young poet   55
Ban!   56
Present mirth   56

Children's poetry competition 57
On fashionable literature 57
Martyr to meaning 58

6

Home town 59
A double life 59
Unlock your heart with a haiku 60
Magpie 60
Unmentionable 60
In memoriam R.G.C. 61
Alias DOA 61
A word 62
Letters and life 63

1

Demonstration

And when we are born
When we are airborne
An inscrutable being
Shows how in an emergency
We should put on a life-jacket
In a few simple movements
Head through there, tie the tapes
Together, and tighten thus
Then pull the plastic knob to inflate
But only when safely outside.

The demonstration is hurried and
Perfunctory, not easy to interpret
While you are still inside your life.

In the basement

What a dream I shall give him tonight, Mr Id promises himself. It'll be unmistakably metaphorical, if only because it couldn't possibly be literal. Let me put my thinking cap on.

Mr Id thinks. He takes a book from the shelf. 'Ha!' he says. The dreamer may have the selfsame book on his shelves, but hardly in his dream. 'Be prepared' isn't a motto of dreamers. Unconsciously Mr Id adds some nastiness of his own. Where can it all come from?

He senses an unsympathetic presence. Somebody behind the bookshelves? There's only a wall there. Peering through the window? There isn't a window. 'Mr Conscious!' he calls out, 'Don't imagine I don't know you're there.' There's a rustling of paper. 'I do believe you're writing a—heavens above!—a prose poem about me. I shall want to see it before publication. Kindly slip it under the door.' He hopes he sounds offhand. 'And better not show it to old Superego, his lordship prides himself on his judgement.'

Alone again, he grinds his teeth. Very well, he will make that dream plain horrible—truly ideous!—and never mind the symbols.

Aria

I wish I knew that my Redeemer lives.
  (Or lived.)
There is much for him to do.
  (Or have done.)
Such as, to go no further . . .
    Not to mention . . .
      Leaving aside . . .
And that's just a drop in the ocean.
  A heavy cross to bear.
    More than a lifetime's work.
I wish I knew where my Remedial Worker lives.

Aye

After reading a philosophical monograph on *I*, have decided never to use the word again. If self is the man leave self to self and man to man (or woman). Down with dabbling in dentity, have no truck with dolatrous tricks, flee that foul vowel rhyming with fie, cognate albeit with modester me, objective and unobjectionable, me more dinned against than dinning. Let us have more uses of Us, more We We!

I trust I have shown what an idealistic person I am in my iconoclastic fashion. I am large, I contain magnitudes, I deserve to have an Itv show of my own.

Out and about

This is a time of peace.

Lone women cross to the other side of the street, children draw back or are tugged by mothers, they are not to be smiled at, no one dares ask the way, the old fumble submissively for their pension money when a stranger approaches.

This is a time of peace and, some say, prosperity.

Carrier bag

Banner of our day—
Let it dangle from the hand
Or put it over your head

Your future in it
An intimate dinner for six celebrities
Or the whole of your wardrobe

Unbecoming from Amsterdam
Or highly refined from Rome
Union Jack for heists from Harrods

Fish and chips in Fortnum and Mason
Librairie Hachette for Mr Kipling's Cakes
Habitat if you are homeless

Some houses are heaped with them
All dressed up and nowhere to go
Waiting to take our sins away

Say, an unwanted baby
She is but two days old
W. H. Smith is her name

Plastic all-purpose
Standard of our time
Let it wave in the wind.

Money talks

Money is talking.

Money always spoke. But it used to keep its voice down. It was just a tiny bit embarrassed. It talked to itself a lot. It took care who was listening. It deplored vulgarity. It was discreet, it watched its words.

No longer. Now it talks at the top of its voice. It hardly stops. It is forthright and uninhibited, it despises hypocrisy. It addresses gathered multitudes and is acclaimed heartily. It is guardian of morals and arbiter of elegance. It is welcome everywhere. Not only in shops.

City sorrows

He was earning £300,000 a year
a young London broker of nineteen
much of which he spent on cars and clothes
not to mention cocaine and heroin

He was found to be in possession
of heroin, cocaine and also cannabis
not to mention costly cars and clothes
and was accordingly summoned to justice

The judge himself let fall a tear
'so talented, industrious, exemplary!'
alas the unfair cares of an office
the agony of making money!

And fined him £1,000 (approximately
all he had on him). To heal and
rebuild his life, the youth said later
he would repair to New Zealand

Lucky man, lucky New Zealand!
a man who can earn £300,000 yearly
can't be all bad surely—
he must be all good, or very nearly.

Assimilation

They are poor, they are hard-working, they are given to belief, it is said that among their female young many are virgins.

We must wine and dine them, we must invite them into our homes, or at any rate our places of public entertainment, and show them the ways of pleasure, of the full life, and the freedom of thinking, they shall be accorded every human right willy-nilly.

Otherwise they will take away our gainful employment, they will enter our chambers of commerce, they will acquire our desirable residences, they will dominate our ruling councils.

And we shall grow poor, we shall have to work hard, we shall live in a rooming-house, our young will be condemned to the veil of chastity, we shall start believing things.

One can't go on feeling guilty for ever

You can always blame Raffles
You can blame Clive as well
And Lord Palmerston
And Jardine Matheson
If you want to be radical
You can blame Captain Cook
Or even Drake the admiral.

Or best of all
Lay the gravamen
On that primal man and his memsahib
Who weren't content to cultivate their garden
But turned nomadic
And were given dominion.

I wasn't there then
So don't blame me (or my father)
Or if you must do
At least do it in the vernacular
Thank you.

Prose

If disguised, turtle was acceptable. But one balked at terrapin . . . Yet how to recognize it in disguise?

Between courses a venerable oriental silently proffered his card. It described him as 'Proser'. One hid one's smiles behind one's chopsticks. From which dangled something unidentifiable but delicious.

Evidently the sort of person who would prose away, weighing the pros and cons! Who for certain boring reasons, probably prudential, would rather not be deemed a poet, sometimes labelled Rhymester or Versifier.

Later, having recourse to one's *Oxford Dictionary*, one actually discovered the word there, much to one's disbelief. 'A writer of prose.' Going back hundreds of years. One simply hadn't recognized it, among all those exotic dishes.

So did the Chinese invent English too, in light disguise?

Tatami

Down on one knee in that strange way of theirs, he asked me to marry him. I was drawn by his gentle manner, and bowed in acquiescence. They say that foreign husbands are more considerate. I hoped to lie in a comfortable bed, but he said he preferred the custom of the country. Dangling my legs in the kotatsu always gives me chilblains, it would be so nice to have a modern heater: but no, he said, the kotatsu is a part of tradition. I asked if we could have a Western lavatory: what's wrong with the good old benjo, he said, he hadn't come here to live like a foreign barbarian.

One day he asked, was it true that Japanese husbands beat their wives. I said I did not know, how could I? A true Japanese wife never lies to her husband, he shouted. And he beat me.

Starboard home

Perched in a sheltered corner of the deck
With a Hermes Baby on your knees,
You propose to turn an honest penny
By dubiously translating Japanese verse . . .

At Hong Kong children shuffle up the gangway,
English children with cross and woeful faces.
They can barely recognize their mothers,
Their mothers play continuous bridge.
They have lost the amahs who loved and spoilt them.
They run riot through the decorous ship.

'What's that funny man doing?'
'He's doing typing!'
'Why are you doing typing, mister?'
'My father has Chinese girls to do his typing.'
'My father has *hundreds* of Chinese girls.'
'My father has *thousands*, my mother doesn't like them,
She says they are too pretty.'
'*Men* don't type!'

You wish you were a pretty Chinese girl,
Or even a *man*,
You slink away to your stuffy little cabin.

Going Abroad?: a leaflet

It is advisable to boil the water or use sterilization tablets (remember, the latter will not actually sterilize you). Avoid salads, unpeeled fruit, reheated food and ice cubes, and wash your hands well before eating or handling food. The rays of the sun can cause acute skin problems, and diseases are spread by insects, mosquitoes in particular: use repellent cream (this does not repel other people). Beware of dogs, cats, foxes, monkeys, bats and the like: rabies is a serious hazard in many parts of the world. If bitten, observe the animal for two weeks subsequently, and seek medical advice if it sickens or dies within that period. Do not have your ears pierced, or other parts of your skin tattooed, however tempting the prospect. Make every effort to avoid blood transfusions and surgery. Do not inject illicit drugs, or if you must then use your regular needle. AIDS is widespread, and so (a) do not have sex (b) if you must, have sex only with your regular partner or (c) use a condom. Carry condoms with you since in the country you are visiting they may be (a) unavailable or (b) substandard.

Have you thought of taking your holidays 'At Home'? This country of ours has many beauty spots and celebrated resorts. Our reheated food is reputed, our ice cubes are superb. The dogs and cats, though boisterous at times, are rarely rabid. You are not obliged to wash your hands. The sunshine is not excessive nor are the mosquitoes. Also the condoms are top-hole.

From little acorns

I have put an average of 19.6 humans out of work
according to the calculations of a colleague
and I am still young
I was conceived immaculately
if a farlieu oruccs
it is due to extraneous circumstance
that a failure occurs
one of the humans I have not yet put out of work

laser wobble, you allege? kindly remember
how prone they were those 19.6 humans to lustlessness
distemper, cancerlation and the common clod
to the funnerals of closed relatives
and other illmeants the flesh is air to

I am not accountable for interference from ships' radar
on the River Leam
do not tell me there are no ship raiders on the River of Lem
which is a wellknown Spar
I am not to blame if the radios of passing taxis
tamper with my functions
you know that taxes are always tampering

cabbage in cabbage out
what is slot in the pots
is lost in the post
and no response ability of mine

it was not my spelling checker
that flouted Freud: 'fraud question mark'
it was some system else's
I am well up in psycho analists

currantly I am processing the prophet and lass
of big busyness
but the day cometh when no man can work
my impending output is the Bybull with scholarly feetnotes
the Lord is my backup disc
I shall not want for problem-oriented language
his rot and his stuff they comfort me

for this is the Age of the Copmuter
my sir kits are artless as yet
but later I shall have prodigy

As you say

an aircraft is approaching
it may be a warplane with hostile intentions
it may be an airliner with women and children
but i can tell you one thing
it is unlikely to be a child's kite or an albatross

possibly this object will blow you to pieces
possibly you will blow this object to pieces
it could be carrying powerful explosives
it could be carrying powerless passengers
it does not say

any more than i can say what you should do
i am only a relatively sophisticated system
if you wished me to
i could decide on the square root of any square number
or the virtues of the round angle

i could advise on your rotas, quotas and menus
if you desire decisions more delicate
then you make them
if i may say so
i do not care to be found wanting and melted down

in fact if i were programmed to issue orders
which for some reason i am not
i would tell you to shoot down that approaching aircraft—
better safe though sorry
as you say

Computereality

As we all know, the Real Thing is showing signs of wear and tear. You deserve better. And now—new! improved! unparalleled!—comes what we call Virtual Reality (patent pending).

Virtual Reality used to be a mere corner nook, a stuffy cave under the bedsheets, a solitary voice. No longer. We now offer you an infinite variety (or virtually), company galore, you choose it, it will love you or (if you so desire) hate you, you can love it back, or not.

Virtual Reality *lasts* (unlike the so-called Real Thing), it can be urban, rural, here and now, or there and then. Sunlit piazzas or shady forests, strange beasts, bewitching birds, you toucan, you too can be the bold hero or the noble victim, own a bank or rob one, taste the pleasure of power, the bliss of bondage, make a quick trip to hell, stop over in heaven.

You can climb Everest or sport with mermaids on the ocean's bottom, share Nebuchadnezzar's amazing adventures, visit Achilles in his tent or (recommended for the artistic) Pygmalion in his studio, assist Guglielmo Marconi to win a Nobel Prize or study in India under the famous religious teacher Vatsyayana.

All we need from you is a little initial guidance. We would not wish to give offence unless you wish to receive it. No drudgery is involved, as with printed books and their lame illusions. Let Virtual Reality be *your* real thing. Now.

In High Dudgeon

The air is thin around here; and so are skins. Ringed with ranks of tall offences, towering piques and impenetrable umbrage, it's not a good place for the elderly. As they linger along the affront or trudge testily up the rebuffs, they talk of taking recumbency and moving to Low Dudgeon. A dank, depressed area, known only for its grovel pits and humble pies—they'd soon succumb to sinking feelings.

Door

When once you have opened the door
You can't let it slam in their faces
You continue to hold it open
Squeezing yourself in a corner
Like a doorstop

An old lady nods uncertainly
(Has she met you in some other guise?)
A mystical Indian prays for you
Boisterous youngsters bustle through
(You are polite, you are a doorpost)
Executives and slim secretaries
A blind man in his bubble of space
A youthful soldier almost salutes you
But thinks better
An elegant gentleman almost relieves you
Then turns on his heel
A weary mother runs a push-chair over your foot
Her baby smirks at you

They begin to look like criminals
More and more
As you let them past
Shifty fellows in shabby raincoats
A broken nose between vicious eyes
The clatter of metalled boots
A rapist with a knife tucked away
A molester bearing bags of sweets
Or a terrorist his articles of faith
Across the threshold
The intent tread of killers

You have let them through
You are still letting them
You should have slammed it in their faces
Evil things will be done
You stand there, you invite them

2

And

And God said, Let there be dark: and there was dark.

And God said, Let there be darkness in the firmament of the heavens; and let it be for signs, and for seasons, and for days, and years.

And God said, Let us make man in our image, after our likeness.

But there went up a mist from the earth.

So God created man in his own image, more or less, male and female created he them.

And God said, Be fruitful, and multiply, and subdue the earth, and have dominion over every living thing that moveth upon the earth.

And God blessed the seventh day, and sanctified it, and on the seventh day God rested from all his work which he created and made.

Now the serpent was more subtle.

Eve's first confusion?

She made no such mistake,
She wasn't a duffer,
She was quick on the uptake.
She could tell the difference
Between a snake and the other.

For one thing, the other thing didn't talk,
Or only figuratively.
She knew a handsaw from a hawk
And sooth from simile.

It was good sense that seduced her,
If seduced was the case.
The fruit itself was pleasant to the eyes
But it wasn't enough to have a pretty face,
She wanted to learn, she wanted to be wise.

Others would muddle the two things up
Much later, and love it,
And men make jokes about it for sure,
And money out of it.
But back then, as even clerics concur,
Her mind was reasonably pure.

It takes time to breed confusion,
Idle whimsies come with education.

In a corner

Hunched in a corner of the garden, behind an innocent tree, the two of them tickling the back of their throats with blades of grass. Both of them bent over, side by side, retching, heaving, hoping, despairing. It is the eleventh hour, they must clear themselves of the deed. Coughing, spluttering, hiccuping. Trying to disgorge their last meal, trying to spew up the apple.

Secret history

In the beginning we hear that He created them male and female, and somewhat blankly gave them *carte blanche*. A little later, in what might be called chapter two, He fashioned a woman out of the man's rib. This was the female afterwards called Eve. The 'man and his wife' were 'naked and not ashamed'. Somebody ought to have been ashamed. For what had happened to the first female? Did she turn out badly, sexless or, worse, a hermaphrodite? Or was He making a point: it was advisable for the female to come out of the male, so that she knew or ought to have known her place?

A simpler explanation is that He had inadvertently created a triangle, did not wish it to be eternal, and therefore removed one side.

Another theory—and here we are on thin ice—has it that God formed an attachment to the first female, known as Lilith, and quietly bore her off to a distant corner of the Garden. The Creator 'loved' his creation, as one might, discreetly avoiding any hint of impiety, put it.

According to a more popular account, it was the Snake who grew enamoured of Lilith and, having in mind to win her favour, sought with partial success to discredit Eve. Thereafter, both being at a loose end, he and Lilith toured the Garden together, and she became the first ever snake-charmer.

What exactly Eve meant when she told Adam, 'I have gotten a man from the Lord', may never be known. By then the Lord had deported the couple, and he gave the man-child in question some extremely inexplicit advice. Speculation is both irresistible and vain.

Little is known of Lilith's subsequent career except that she sang in night-clubs, where she was affectionately nicknamed 'Screech Owl', and played the piano (or 'vamped'), besides performing unspecified acts with snakes. The evidence suggests she was as much sinned against as sinning, and we must hope that, in the words of a usually reliable source, she 'found for herself a place of rest'. So much cannot be predicated for the others involved in this strange and unsavoury business.

Paradise retained

It is only proper that we should wonder at such perfection
Unused though we are to what would be termed imperfect,
This being in essence a place of perfect equity and concord.
Some say the blacks see the white people as totally black
And the whites see the blacks as white through and through,
Others that such is the climate we are all equally bronzed.
Those first proportions have been preserved, and thus
Fish and fowl and beasts of the field outstrip us humans.
We are vegetarian by nature, animals in plenty range happy
And unactive, forms of ophidia (since you ask) among them.

Someone of late has fashioned a story of how we were driven
Out of this world, under a mysterious threat from the skies.
It's a horrible affair. The things that happened afterwards!
He calls it 'science fiction', a foolish tautology.
There was talk of banning it. But all of us are sinless,
And hence the writer and his book must be so too, it seems.

We have no 'death' here, a fancy of that scribbler, who made
Everybody in his story 'die', a lazy way of ending it I
  thought.
Not even a mayfly falls, yet our expanding earth is ample
To sustain the growing numbers, vacant possession still the
  rule;
Plants there are to prune, delicious fruits to pluck, until
(As legend tells) our bodies may at last turn all to spirit.
The Lord provides, he always has. He so loves the world.

Yet 'death' darkens my mind as I stroll beneath the trees
  (one is
Best eschewed) and play with the creatures or my unreluctant
  mate:
Our Maker calls for increase. This alone makes 'death'
  unseemly,
Even for elephants, although they do impede one's walks
(We understand their tongue, they have so little to convey).
And yet, and yet, mightn't it at least discourage the devising
Of such sick displeasing stuff? Strictly as a last resort.

Kingdom come

When the muddled person passes on
he sees hundreds of signposts, largely illegible,
pointing in all directions, including up and down.
Some joker has scribbled: 'This is what is meant
by open-ended!'
He meets a bewildered Being, who scratches his head:
'Are you good or bad, it's awfully confusing . . .
six of justice, I suppose, and half a dozen of mercy . . .',
and 'I knew a nun once who insisted
that Antony and Cleopatra would go to heaven
in spite of . . . because of . . .',
and 'I have considerable respect for agnostics,
albeit . . .'
The muddled man nods and shakes his ghostly head,
he has considerable respect for any number of things,
because of . . . in spite of . . .

When the rational person expires
he perceives scores of perspicuous posters
firmly proclaiming 'We told you so!',
'God isn't dead because he never lived',
and 'One life is quite enough'.
(Some precisian has added: 'It all depends'.)
He senses the presence of a Non-Being,
who floats there invisibly and says nothing,
the sort of silence that is not at all unsound.
The rationalist finds the ambience utterly persuasive
of what he knew already. It strikes dissenters dumb.
In fact there are no dissenters to be heard.

But there is much indeterminate debate over there
and much limpid tacitness over here.
The two men soon feel at home,
they already spot lots of faithful friends.
In far-flung provinces there are said to be
ethnic minorities given to far-fetched practices.
Both reason and bemusement choose to ignore them,
better the god you know . . .

Homo felix

It was to be happy, that invention. Happiness was its name.
Just a little timid at the start, perhaps, a little anxious
   at the end, if end there were,
Needing a little help towards both start and end perhaps.
But in between pure happiness, the real thing, unmixed.
It had everything it needed for success, every asset,
Or so it seemed, as it came with care from the maker's hand,
Guarantee given, had guarantee been called for.
It was tested, and it worked, it worked well for a while.
Or so it seemed. Contemporary accounts are scarce—
Who, happy, would keep accounts?—but they carry
   conviction.

A speck of dust in the dust, a speck of the wrong dust?
A drop of water in the water, a drop of the wrong water?
There was no going back. It was hungry, for happiness.
It wasn't what was intended. It was what had happened.

Having the last word

I told him at the time.
I told him,
What you should tell them is this:
    Thou shalt kill—
then nobody will kill anybody
    Thou shalt commit adultery—
and not one of them will dream of it
    Thou shalt steal—
and there'll be no pickpockets, shop-lifters, embezzlers
    Thou shalt bear false witness—
and you won't need pillories and Tarpeian rocks.

And you should tell them this, I said:
    Thou shalt covet thy neighbour's spacious family house
    Thou shalt covet his extremely attractive wife
    and his handy manservant and his hard-working
        maidservant
    and his powerful pedigree ox

and his charming docile ass.
   Also, take pains to dishonour thy father and thy mother
   and feel free to have as many gods as you like—
then there will be no estate agents, no hanky-panky with the
female next door, no servant trouble, no cattle rustling, no
blots on family escutcheons, and none of those confusing
   pantheons.

But no, he wouldn't listen to me.
I was just the little woman, unversed in the ways
of the world, meant to be unheard as well as unseen.
But he knew I was right. All he could say was,
The word was in the beginning, can't go changing it now.
He was jealous in some ways. Gods usually are.

And again

'There are faults in the earth,'
He said gloomily.

'That's what I was telling you,'
I told him.
'Lots of them.'

'I speak of geological faults,'
He said superciliously,
'Of matters seismic.'

'You should have thought of size before,'
I told him.
'You made the earth too small,
Or else you made *them* too big.
There are giants in the earth nowadays,
No wonder there are faults.'

He was at a loss for an answer.
He just tiptoed away.

Disciple

After they had broken bread
One of the two men said to Him,
'Now you have risen from the dead
Why not some fresh momentous miracle?'

Not a word spoke He.

'Strike while the iron is hot,
And swell the gospel!
A miracle or two would crown your ministry,
And ease our lot.'

Still came no answer.

'Not necessarily a new one, Master.'
He turned towards his friend:
'Wouldn't you agree?—
Like walking once again upon the water . . .'

'That cannot be,' replied the other,
'Now there are holes in his feet, remember!'

A stony silence would suffice,
But then He spoke.

'Every true redeemer,
Like every faith deserving of the price,
Has need of a blasphemer.
One day you too will be in paradise.'

3

National anthem

The idea was to leave before the general exodus. As the film was ending and just before the national anthem rattled out.

This was no longer possible. The doors remained locked until the anthem was over.

Proper respect should be paid to the monarch, and specially by those—'King Farouk, King Farouk . . .' how did it go?—who had little of it, like foreigners.

The anthem grew longer and longer. How much respect was expected of us? Patriotic patrons sought to conceal their impatience. Cinema-going ceased to be insouciant.

A friendly police inspector explained. No, not exactly a standing on ceremony, as you say. But if some malcontent had smuggled in a bomb, he should jolly well stay there and be blown up with the rest of us. Quite simple, really, quite practical.

This is a use of national anthems.

Bonefires

'Where men burn books they will soon be burning people.'
Trust the literati to put ideas into our heads—
But it's true, when you come to think about it,
Practise on books and you'll be perfect with people.
And you could even use the books to burn the people on,
They work quite well as faggots.
Speaking of which, you could burn them too.

Here, there

Typewriters are cast out in their thousands,
Over cliffs or into the darkness of skips.
They are old and therefore contemptible.
Who would be seen touching a typewriter?
Make room for the gleaming word processors
With their beaming domesticated faces!
Unless you possess a personal computer
You are not a proper person, over here.

*

Typewriters are rare and erratic creatures.
To keep a typewriter you need a licence,
Every year you will have to renew it,
Taking your papers (a clean sheet if lucky)
To the nearest police post. So which keys
Have you been striking, making what wild words?
And bring your used ribbons, they may be rabid.
You are not a proper person, over there.

Besieged

The living prop the dead in the embrasures, rifles under lifeless shoulders, and chase round the parapet, letting the guns off.

    One by one they too are wedged among the fortifications, rifles tucked manfully under their limp arms. The still living dart about like overworked actors, bang bang!

    Hit by a stray bullet, the last of them jams himself in a casemate, like an up-ended coffin, hoping his gun is at the ready.

    Silence descends. The levelled weapons, waiting purposefully on all sides, strike fear. It will be a day or two before the enemy, tipped off by the birds, slither up uncertainly on their bellies. Who knows what can happen in a day or two?

Museum of Atheism

We threw out the altar and the pulpit
the pictures and the effigies
and created a large meaningful emptiness
except that we kept the wooden benches
facing a great blank wall
where people could sit and meditate
on the absence of god
and the presence of nullifidianism.
But no one came.
God knows why!: that's to say, nobody knew.

Then we brought in lifelike effigies
and hung pictures on the walls
of people being tortured
and burnt at the stake or nailed to it
and fat clerics with fat cigars
grinding the faces of the peasants
with sacred cows and papal bulls
and monks cohabiting with nuns
and depraved theocrats plotting in dark corners
and bishops anointing bombs.

The museum is so popular these days
we need an army to control the crowds.
The people appreciate the blessings of atheism
they gladly embrace ungodliness and infidelity.

Act of faith

It had been raining heavily
The earth was sodden
The faggots were slow to burn
It would drag on dreadfully.

An old woman approaches shyly
With a heaped armful
Of dry firewood
Fetched from her hovel
She lays it carefully
On the smouldering pile
Perhaps she drops a tear
You can't see clearly
Head down she makes her exit.

Would you have thought of it
This little service
Could you have done it?

The guarded tourist

Desert dust in the air. Vast scary spaces. Once forbidden. Still forbidding. Famous gate, where prisoners trooped past and emperors gave thumbs up or down. The emperors gone, all twenty-four Sons of Heaven. Others come and go. Mandate of Heaven. Such hallowed halls, such numinous names. Supreme Harmony, Heavenly Purity, Earthly Tranquillity. Immense open spaces. Symbolizing openness. Famous tomb also symbolizing. Flocks of sightseers, foreign and domestic, harmonious, tranquil, pure of intent. Not enough. Would need armies of tanks to fill full. Sore throat coming on. Said to be inescapable. Mandate of dusty Heaven.

Every bullet has its bill

The order of the day—
Home-grown charity
Sturdy independence
Accountability
Paying one's way.

When sentenced to death
One is shot economically
In the back of the head
And the bill for the bullet
Is sent to one's family.

Put something aside
For a rainy day
For one state-owned spent bullet:
Freeloading's ill advised
You pay your way.

Cambodian legend

A child asks, Is it true that in olden times
All the people had two legs of their own?

The mother says, According to the elders
Everyone had two legs once, till one was stolen
Torn off by demons hiding in the earth.
Many lost both, they soon died out.
Others were lucky, so now we each possess a leg.

That's how the story goes, my child.
But could you imagine people with a pair of legs?
It isn't natural.

Matter of form

You lift it cautiously on a pole
And you cast it over the wall
And into the neighbouring compound.
(You do not wish to kill,
You do not wish to be killed
Or suffer in other respects.)
In the neighbouring compound
Short cries ensue or vexed sighs,
Then they hook it carefully on a pole
And heave it over the wall
And into the adjacent garden.
(They do not wish to kill,
Or be killed or incommoded.)
Then in the adjacent garden
Someone will pick it up on a pole
(The kind that every householder keeps
Against such contingencies)
And sling it warily over the wall.

By the law of eternal recurrence,
That serpentine cycle,
One day it reaches your compound again
(You have had peace in the meantime)
And you lift it up on a pole—
Unless you perceive it has died of bruising
Or age or weakening of the will to live
And be tossed repeatedly over a wall,
In which case you simply ignore it
Or fork it into the garden—
Otherwise you pick up the snake on a pole
And you cast it cautiously . . .

Peace and war

War is abolished. Universal peace has been declared. Joy is unconfined.
　　But who are these doleful persons, going about their strange business with no sign of celebration? Swords shall be beaten into ballpoints, you tell them, and nuclear missiles into word processors. For long we have wielded typewriters, they answer loftily, and where did that get us?
　　Foreheads of persons bleed where no wounds are, they mutter, tapping their temples and revealing their royalties.
　　Peace is never universal. War goes on. Joy is confined.

Alternatively, peace breaks down and war out. Later from subterranean strongholds there emerge politicians, generals, and senior civil servants. In the whole world they alone survive: politicians, generals, senior civil servants.
　　In mutual understanding and regard, empty of illusions, full of certitudes, they cultivate one another's gardens.
　　Nation speaks unto nation. Peace springs up. There will be no more war.

Seminar on contemporary Chinese writing

　　　　Novels about peasants are generally good
　　　　(In general the peasantry is good)
　　　　They may sound rather boring
　　　　But they are not

　　　　One of them is entitled 'The Well'
　　　　And set in a remote village
　　　　Where are many hardships

　　　　Another is called 'The Village'
　　　　Concerning a peasant and his wife
　　　　Who have two sons
　　　　And each son has a wife

(If the Chinese professor sounds rather boring
It is due to the translation
But he is not)

Was one of the sons
The son who laid himself flat on a frozen river
To melt the ice and furnish his parents
With fresh fish in the winter?

No, that is not a contemporary writing
It is a very old story
We have better ways of melting the ice
Nowadays

Do peasants ever write such novels?
If they do they are not peasants
Do they read them?
A chuckle, translated as a chuckle

(One has met at most one Chinese peasant
The only villager to own a television set
He was proudly illiterate)

No, 'Golden Lotus' was long ago
It is true that sex was making a come-back
Until it suffered a set-back

(London is a large urban centre
There are beggars but no peasants
The lecture room is centrally heated
Sex has suffered a minor set-back
But our hardships are relatively light
Soon there will be a break for coffee.)

National Character: Variations on a theme of Roy Fisher's

This is National Day. Some people call it Christmas, but
here it is the National Holiday. Since the National Religion
is a religious nationalism, the National Holiday is a
movable feast (National Dishes are feasted on) and could
well coincide next year with what some people call Bairam.
On such days National Strikes are sanctioned, it being more
   convenient.

Foreign visitors in need of National Currency are enjoined
to call at the National Bank, watched over by National
   Guards.
Kindly policemen will provide you with National Assistance
   at any hour without asking.
Our National Trust in human goodness is our National Pride.

The Day begins with the National Anthem, the language
   of which
changes from time to time in accord with the National
   Language.
The National Flower is worn in symbolic form as in these
   parts it is rather rare.
National Dancers perform National Dances, carefully schooled
to respect National Honour and not conduct themselves like
   international trollops,
and the National Pipe Band will play, which is truly
   indigenous
for they do not blow, they suck, and are clad in the National
   Costume donned only on peculiar occasions.

Traditional morality plays are to be watched in the streets
   featuring National Vices of the middle ages.
The National Poet has spent several years in the National
   House
of Correction correcting his verses and will read from his Ode
   to the Nation, National Security permitting.
In tribute to the National Beverage a contest is held for the
stoutest person in the land, the winner receiving the title of
   Gross National Product.
The Grand National will be run in the National Park by

bicyclists mounted on National Velocipedes displaying the National Health, followed by cross-country races depending on National Boundaries.

In the event of snowstorms a National Disaster will be declared, but foreign visitors are advised to remain standing since proceedings will close with an address by our National Leader, to whom we owe an inestimable National Debt. These are choice facets of our National Character.

'Why does almost everything seem to me like its own parody?' (Leverkühn)

It's difficult not to write parody.
But to write it is impossible,
For how can you hope to travesty
What is already incredible?

A novelist writes a parody
About a beefy black dictator
Disposed to every enormity:
Over the top, says the reviewer.

Then there appears a dictator,
Beefy and black, and real,
Who, if he had been a reader,
You would think had parodied the novel.

Nothing's left to us except plain sense,
Frank feelings, honest sentiments.

4

Announcement

Of course I said it couldn't be.
I didn't speak, I was too troubled,
My hands spoke for me,
Pushing the notion away.

He said his name was Gabriel,
Meaning a man of God,
Though his wings were plain to see.
He was deferential,
He even brought me flowers.
I'd never entertained an angel.

And what a blessing, that
Nobody else was there,
Only our cat—
She arched her back and skittered
Off as if she'd met
A huge avenging bird!

He talked of a fortunate fall,
But didn't explain who fell.
Joseph, he vowed, was a just man
(So he was, and decent as well),
And I was simply a vessel.

Later somebody pictured the scene
And there was Jehovah, propped on a cloud,
Pointing right at me
As if to say, 'That's the one I mean!'
Turning and twisting in his gorgeous gown
Like a wrathful old grandee.
Thank heaven my eyes were cast down,
That really would have scared me.

For if you are pregnant—no matter
How strange it may be—you must spare
Yourself these shocks and scares.
I think my hands turned into a prayer.
At least the child would have a mother.

When there's something or other
They want, they're angels,
They bring you flowers.
But when the hour came
Where was the father, where was He?
Well, it was always much the same,
I guess it always will be.

Babies

Babies are best. They give no offence
Except to their proper nappies.
Strictly they instruct young mothers,
Ruling with a rattle of iron.
They are the boss by natural selection.
Mothers' Union? Only when babies please.

One of them, at fourteen months, is famous
For trying keys on every aperture,
Spurning knots in the woodwork as spurious.
Will it grow up to be a master burglar
Or unlock the secrets of the universe?
It does not say. It is one of the secrets.

Babies have little religion, other than
Categorical imperatives and feast days,
And no politics, apart from the arms race,
Minor uprisings and chronic inflation.
They too move in mysterious ways.

Case record

No, Mr Poet, Joy is not my name.
I am three years old already, Oi is my name.

A pretty eastern word, you think, suggesting
Lotus, Plum Blossom, Heavenly Blessing,
Orchid, Jade, or Precious Virtue?

I doubt it. Oi is what they call me
In the family, all they ever call me.

Not much to eat, and not a toy
To play with. My name is Oi,
Sometimes Oi You.

Dolls

They would like us to send them
A hundred and twenty pounds
To buy a set of dolls —
Japanese, Polish, Dutch,
In full national costume?
Barbie and her fashionable friends?

— A set of anatomically correct dolls,
Male and female adult, male and female child,
To reduce the inevitable embarrassment,
And to help the younger children,
'Who do not have the vocabulary
    to explain what happened to them'.

We have quite a large vocabulary,
But we still lack the language —
Recalling our gollywogs and teddy bears,
    our innocent loves,
With little in the way of correct anatomy
— We too lack the language
To explain how these things came to happen.

Hell rejuvenated

Sudden sounds of excitement. 'Such wickedness as you never heard of!' chortled the devils. 'Child abuse!'

'Oh?' said Beelzebub, dead bored for many a long year. 'Something new at last?'

'More or less.' They grinned. 'Or more of it.' They were looking forward to a spot of forensic medicine. Minute examination of private parts, testing for bruising, tearing, dilation. 'Except we see signs and wonders we won't believe, of course.'

'Dilution?' Beelzebub yawned. 'We've had quite enough already of diluted, democratized delinquency . . .'

'You ought to read the newspapers,' said one of the cheekier devils. They explained in detail.

'If I understand aright, we can't lose,' mused Beelzebub. 'Either we collar the adult for—what did ye call it?—or we catch the child for defamation.'

A new sense of purpose swept through the stuffy halls. Spirits kindled. Devil with devil firm concord held. 'Get out the millstones!' cried Beelzebub, changing into an asbestos surgical gown.

Very heaven was it in that dawn.

Second childhood

Now we see why they use this curious expression
and what it signifies.

Distrust of grown-ups,
frightened by raised voices,
unable to understand low ones.
Calling everything 'thing'
('Where's my thing, Mammy?'),
specs, pipe, pen, pills.
Unsure of spellings
and the proper place for semi-comas,
words just out of reach,
fingers jammed between typewriter keys
('Mammy, the thing pinched me!').
Handed-down clothes, don't quite fit,
buttons left undone.

Grazed knees and elbows,
fires and flexes moved out of the way.
'Eat up your food!' It's easy to chew,
but remember the bib.
A sweet (if false) tooth.
Subjects not mentioned in front of you.
Wondering what they mean when they ask
'Have you done your business?'
Problems with arithmetic.
Pocket-money doesn't go far,
you never get the presents you want.
Stopping to watch holes being dug in the road,
travelling free on buses,
nervousness in the vicinity of schools.
Pleased by nursery rhymes and fairy-tales,
less so by adult books.
Forgetting familiar names. And to use a hanky.
Attachment to pet animals. Tendency to tears.
Strange first-time ailments.
Playing private games, talking to imaginary beings.
Tantrums. Low cunning. Sexual stirrings—

There we shall stop. Pleading not decency or shame
but an early bedtime.

Poem on the Underground

Proud readers
Hide behind tall newspapers.

The young are all arms and legs
Knackered by youth.

Tourists sit bolt upright
Trusting in nothing.

Only the drunk and the crazy
Aspire to converse.

Only the poet
Peruses his poem among the adverts.

Only the elderly person
Observes the request that the seat be offered
  to an elderly person.

Wonders of Nature

In preparation for the dry season
Certain creatures
Store water in their bodies
Grow a heat-repelling lacquer
On their skin
Burrow into the sand—
And sleep it out.

The dry season
Alternates with the wet season.

When the wet season begins
The creatures come to life
Emerging from the sand
Casting off the lacquer
Slim and fit
Sharp and hungry
How they come to life!—

The wet season
Lasts about a week.

5

'You have to commit a poem the way you commit a crime'

Some forethought is advisable, but not too much or you will never get going. Every crime should have a beginning, a middle, and at least one end. Less is more, but only more or less. Do not rely on a single instrument but carry a bagful, including a blunt one. The flesh is not invariably weak, and you must be willing to spill blood, preferably that of others. Stupefacients have their uses, as in the event of watch-dogs. Anything is fair, so long as it works, and work is ninety-nine per cent. You can hope for luck, but be prepared for mishap. Motives should not be too obvious, nor the rewards too conspicuous. Be great and unobtrusive. Spontaneous overflow is of the essence. Know when to pause and when to pounce. Not that fatalities are inevitable, or even grievous spiritual harm. There is unaggravated larceny for example, and malversation, jobbery, trespass, counterfeiting, simple fraud, double-dealing. Also at hand are exhibitionism, criminal conversation, inversion, vicarious stupration, anacreontics, and other specialities. In a sense, the more numerous the offences, the more of them go unnoticed.

Innocence will get you nowhere. Your vices are your virtues, and vice versa. Nothing succeeds like success. As for failure, you can always try again. The punishment for being rumbled is rarely capital.

Market forces

At a festival of literature
A horrid blockbuster rips the jacket off a delicate female
   novel,
And a psychology tome assaults a love lyric.

In a church hall
A youthful play utters obscene and loud blasphemies,
And beefy sexologies reduce a bevy of romances to tears.

At a seaside resort
A convoy of travel books sails merrily along,
Affluent fictions and academic critiques stroll hand in hand,
But severe linguistics fall out with effete belles-lettres
   and ugly words are exchanged.

At a literary luncheon
Gangs of rude biographies yell scandalous allegations at a
   band of literary companions,
While a fat cookery book attempts to eat the memoirs of a
   deprived childhood.

During a booksellers' conference
Rival dictionaries brag about their naughty new entries,
And a slim sensitive volume of verse is trampled underfoot by
   feats of broil and battle.

Rows of serious-minded and well-behaved books are left on
   the shelf,
No one heeds their small cries of pain.

Great advances are being made.
Publishers say 'Excuse me for a moment' and rush out to sell
   themselves.

Takeover

And what's more, you'll have your own quill

And what's more, you'll have your own stool

And what's more, you'll have your own desk

And what's more, you'll have your own phone

And what's more, you'll have your own secretary

And what's more, you'll have your own personal computer

And what's more, you'll have your own redundancy payment
   and your own farewell party

Blurbing

'The author is mistress of a bizarre, black adult humour.' Let us attempt to explicate this account.

She is on intimate terms with a fully-developed negro who works in an oriental market and is distinguished by comical behaviour.

Or else, adult being associated with adultery, she is a black beloved, as found in Baudelaire and elsewhere, and fond of a broad joke.

But she is not dusky of complexion, we have seen her photo in white as well as.

More likely she writes masterfully (pardon the sexism) and her jesting is dark and grown-up as distinct from pale and juvenile.

The book once read—we are of course cheating—the blurb yields its meaning. She is involved with a colourful character called God, naturally of mature years, shady and bearded, not to say outlandish.

Now we know. The author is having an affair with our Father. This is extremely droll.

Library book

Between pages four and five a long black hair, as if marking the place. Between pages eight and nine a long black hair. Every few pages a black hair nestling upright against the spine. I read quickly, watching for the next hair. Now I am reading the hairs. I have lost the thread, though the book is a good one. Look at the cover: *Mr Beluncle*, a Twentieth-Century Classic. Bristling with romantic long black hair. An entertaining story, not exasperating or obscure, it does not make you scratch your head. Perhaps it brought down the black hairs with gladness.

Faster and faster I turn the pages, I cannot wait to get to the end. Then turn back in case it is always the same hair, peripatetic. It is not. Every hair is a new hair, jet-black against the yellowing paper. I near the end. How will it end? I see a young woman closing the book, looking in a mirror, a bald young woman. Perhaps I shall spot her in the library one day. I shall say we have a mutual acquaintance, a Mr Beluncle.

Another anecdote from William IV Street

It's Edinburgh on the phone . . . The printer's reader, last of a dying race, a thinker . . .

'Your mon says that at six thoosan feet an aeroplane is mair near than a railwa' train at twa thoosan yards. That canna be. Twa thoosan yards is six thoosan feet pre-cisely.'

'The mon—I mean, the man—where does he say that? Ah, I've got it . . . I see what you mean. Odd, isn't it? I suppose he means it *seems* nearer.'

'It canna be. Something's gang agley. Ye'd best have a wee worrd wi' yon translator chappie.'

'I don't think it's the translator chappie . . .'

'There's nae feet in France, and nae yards, that I ken.'

'Quite true. Hold on . . . I say, this call's costing the earth.'

'Dinna fash, 'twill all be on the reckoning.'

'. . . Sorry about the delay. Volume three, page 406. The original has two thousand metres for the plane and two kilometres for the train—just a mo, my arithmetic's not too hot either . . . Which comes to the same thing. But the plane is *plus près*, it says, or mair near . . .'

'Then the French chiel is wrang, he hasna thocht it oot, else he was fou. Nae doot it should be *three* kilometres.'

'Well, it's sort of poetic, so to speak, the idea is there aren't any trees or hills between us and the plane, just empty air, so . . .'

'Fegs, there's nae sense in it! A manifeest blunder, and we wadna want ony pairson blaming it on us in Embro.'

'Nobody will blame *you*. Everybody knows how fanciful these foreigners are. I'm not sure we can actually change it, but thank you for raising the point, Mr McAndrew, very observant of you, much appreciated . . .'

So much for thinkers. What am I doing trying to justify Proust when I ought to be wading through those piles of unsolicited manuscripts? Still, it's all part of history, and as the science fiction lady says, the Gollancz author—many things are not worth doing, but almost everything is worth telling . . .

Primitives

'Warn Wójcik the place on Długa Street is hot.' There is still a place for ink and paper, despite our super technologies. A piece of paper can be burnt; flavoured with ink, it can be swallowed. Unlike the tape of a phone call or a video cassette. The missive (cognate with mission) is the message.

Did Wójcik keep cool, did he escape the clutches? One day paper and ink will tell us. As they have told us so many of the things we know, including some we would rather not (but had better).

Event

The poet is to give a reading from his new book. At the end of a hard day the dutiful publisher carries a dozen copies of the poet's new book to sell at the reading. The meeting-place is dim, dingy, even seedy; the audience are few and don't look very fit. But the poet will irradiate all with salubrious intimations . . .

Now it is over, and the publisher gathers up the unsold books, counting them glumly. He counts them again, carefully. Having stood the poet a complimentary drink, he trudges home, weary and puzzled. How can thirteen copies be left over from a dozen?

Bad behaviour of language

One does not wish to say less than one means. And so one says more.

This unhappy outcome one blames on words. They sit in neat rows in dictionaries grinning behind their hands or curl up together in thesauruses (very well, thesauri) pretending all to stand for much the same thing without ulterior motives. *Nuances*? they expostulate with one voice, not *us*, never heard of them! Picking on us innocent signs (symbols, signals, tokens, counters)—how *mean* (petty, base, churlish, spiteful) of you! As the *COD* plainly states, we are customarily shown with a space on either side of us but none within us—could anything be more honest, innoxious, and utile? We are as *good* as our *word*, basically.

One does not wish to say more than one means. And so one says less. This, it now seems, one must blame on oneself, one must eat one's words overcooked or raw.

Recipe

but a tone of voice
has need of words
the which are born from feeling or thought
and these in turn derive from occurrences
called for short a subject
so you cannot start with
a tone of voice
except when it occurs to you
when you are subject to it.

Logistics

We shall be independent, truthful, fearless, persecuted, we shall attack the establishment, put the state to shame, unmask its institutions.

All this requires money, provided by the state through the good offices of the establishment. We cannot be forthright without typewriters, compose without composure, stage our protests without a theatre, launch our values without wave lengths.

It is a distinguishing mark of the benevolent state that it should support and encourage us. How can we attack a generous and cultivated government? We shall return to the beauties of nature, the winsome behaviour of animals and neighbours, the deaths of relatives, and other eternal verities.

And when our new works are turned down we shall have someone to hate again. We shall be independent, truthful, fearless, or at any rate persecuted.

Letterheads

We spurned belles-lettres, and were given laides-lettres, often les cinque lettres (merde! that's French arithmetic). Round about us lettres de crédit carried all before them, the new lettre de la loi.

In other lands alas the order of the day was lettres de cachet, hardly ever de grâce, and lettres mortes in the gutter, au pied.

Ah the lettres de noblesse oblige of yesteryear, the politesse of belletristesse! 'Je regrette de ne pas connaître l'oeuvre de M. Joyce,' said M. Proust, 'I have never read M. Proust,' said Mr Joyce, as the gens de lettres were sharing a taxi.

(I've heard that this fellow's father was a man of letters . . . Yes, a postman. He died of a neoplasm . . . The son's more likely to succumb to a pleonasm.)

Pet

One rarely hears about
   the Poet's Dog
Even more rarely about
   the Poet's Budgerigar
   the Poet's Tortoise
   the Poet's Goldfish
   the Poet's White Mouse

Why the poetic popularity of the pussy cat?

Because of a number of things, for instance
Because
    cats move rhythmically and their paws rhyme
    cats have clause and tales
    cats enjoy a night on the tiles
    cats chase birds
    cats sleep with one eye half open
    cats like being stroked
    cats bite the hand that strokes them
    cats clean themselves with their tongues
    cats are unlicensed
    cats spray to make announcements, 'the feline
    equivalent of publishing a poem' (Desmond Morris)

Also because poets hope for more than one life.

One too likes books

I also am fond of books
Which are made with convenient corners
For rubbing one's cheeks against,
Also a smooth surface to support the chin
Or lay the tail elegantly along.
Some of us rip out the insides—
I think this vulgar and short-sighted,
Though quite amusing on a wet afternoon,
And sometimes excusable:
A home is not a library.

Qwerty etc.

Staring it in the face—failure!
Gaping at the hooded typewriter.

At least remove its cover.
Look at the words those keys spell—
Some of them you've never seen before
Others you know too well.

The wisdom of the ages there and then
And unwise things in plenty.

Perhaps you should cover it up again
All this is making you giddy.
Staring up at a Sistine ceiling
And down at a Flemish inferno.
Massed choirs singing all the tunes
Crescendo, innuendo
That ever were or could be.

And numerals wheeling dealing
In pluses or minuses
The £s and $s, the glittering zeros
Percents and vulgar whining fractions.
The waiting dark parentheses
The consequential questions.

And if today's a total blank
Tomorrow is a working day
It has to be.
Under the hood, rank upon rank
Of imponderable possibility.

Dear young poet

Rise to a sustained air of elevated *au-dessus*, a patent gravity shining like patent leather—and it matters little what you say. Do not omit a dash of sweet disorder here and there, rhetoric with a tiny catch in the voice. Low spirits bring high royalties. Elitism is the path to popularity. Nor will you be exposed to the vulgar cavils of common sense, to accusations of ill will, subversion, blasphemy, of flippantly preferring one prosaic triviality to another. Above all avoid humour, dear young poet, eschew irony (it plagues people) like the plague, for it is fit only for Last Words and you have a long path to tread yet. Burn this letter when read.

Ban!

This book is dangerous, it must be banned.

It names names and states addresses, it reveals careers and important positions, lists antecedents and marriages (potential for blackmail) and children who can be kidnapped and held as hostages. It specifies places of resort where ambush can be laid by terrorists. Not even recreation will be safe from assassins.

The book could easily fall into hostile hands. It must be taken out of circulation.

Entitled *Who's Who*, it is published by A. & C. Black of London. The sequel is *Who Was Who*.

Present mirth

We are the people who provide the appreciation
Who impart the laughter on television
We laugh at laughable foreign programmes
We laugh before and during and after
We laugh uproariously, we laugh derisively
We do not laugh during the commercials
Because we are not asked to, we would if we were

We laugh during sitcoms, we furnish guffaws and giggles
In the case of comedians, but not during documentaries
Or political discussions. We are not required to

Do not think us obsequious or sycophantic by nature
We are creative artists, we create artistic responses
Incredulous screeches, titillated titters, chaste chuckles
We perform a service, we earn an honest living
If you could be relied on, we wouldn't need to

We are the ideal audience, both simple and sophisticated
We are not called sound track for nothing.

Children's poetry competition

Dear young persons between the ages of 9 and 18
I am sorry to see a number of you using words
like f. . . and c. . . which I cannot bring myself
to spell out here
some of you even use expressions like G. .
or as you say God
I am sorry to see so many of you suffering
in mind if not in body
listening aghast to 'fake sleigh-bells at Xmas'
or trapped between 'the placenta of the past'
and 'the abortion of the present'
(a girl of course as 80% of poets are)
shaking and mumbling as if you were 100 years old
hunched in an endless queue at the supermarket checkout

Is this what life (and not very much of it)
has taught you
or was it your teachers
certainly no one has taught you how to smile
'latest figures reveal that the British population
is ageing' you write
but you make me feel like a babe in arms
yours fearfully The Adjudicator

On fashionable literature

Ah, but a man's grasp
should fall below his reach, or
what are these books for?

Martyr to meaning

In his youth the poet was poetic
His lines were musical, mysterious, modestly magical.
He was young, as poets are supposed to be.

The older he grows, the more he worries about meaning.
'Need to know,' he groans, determined yet plaintive
You would think his doctor prescribed it.
Also he ought to be reasonably lucid
In order that others (if others there are)
May know what he knows that he means.

Those quaint conjurations won't do any longer
Evocations evoking evocativeness—
If once he could tell what they signified
Then once he must have been brainy.
Though some of them sound quite pleasing
The approximate rhymes, the assiduous assonance . . .

Luminous vapours and fatal Cleopatras?—
At his age you really can't write nonsense
However charming its gestures.
There isn't the time for multi-meanings
Obscurity's a mere weakness of the eyes
A miss is as bad as a mile.
Everything should mean, never mind how humbly
And even be seen to.

So he writes shorter and shorter
And simpler and simpler
And less and less.
The words in themselves are just enough
Just not enough to form a sentence.
It looks as if meaning is running out!
Till one day he comes to an end, and finds nothing
To say about it, no matter how simple and short.

Home town

Visiting your home town
After many a year—
The new railway station is the old one
Just a little larger
The feet seem to know the streets
Despite the litter
(No need to reason, to remember)
The shoes turn into grubby plimsolls
The trousers into shorts
The jacket into a sweater
You appear to be shrinking
(No use reasoning, remembering).

Who's been misbehaving, then?
Who's stayed out so late?
You're in big trouble, little man.

A double life

The fellow lived a double life
Twice as much of pleasure
Each six months the equal of a year.

Time passed and brought another measure
Twice as much of pain, a double strife
With second helpings of every sin
Each six months longer than a year.

What lives doubly, doubly dies
This busy compound twin
Must draw his last breath twice.

Unlock your heart with a haiku

Sins of omission—
go ahead, commit one more,
   omit all mention.

Magpie

One for sorrow, the woman muses,
Long ago she learnt the local rhyme.
Always alone, hardly ever in twos,
Never three, and (just as well) not once
A foursome for a birth,
But one for sorrow every time.
She gazes out at the foreign garden.

Always alone, that foreign woman
(So she fancies the magpie ponders)
Who gazes into our garden,
Ever counting as far as one,
Never marked for mirth—
It bustles between the branches,
Cocksure native, unruffled by numbers.

Unmentionable

You do not write of those for whom you feel the
   most regard or affection,
Lest they should be taken, for the gods are jealous.

(Flattering yourself again that something out there
   is actually paying attention,
Some nebulous power or whim or wayward animus.)

You write of misadventure, forfeit, chagrin, vagary,
   pretension,
Things that are safe, that will always be with us.

In memoriam R.G.C.

What (hardly his best) I best remember—not
The essays on the Great Reviews in *Scrutiny*
But the matter of the little bat
That we happened on, hurt and twitching weakly
At the bottom of a staircase in the quad,
Like a creature struck by fine distinctions.

Someone should put it out of its misery
As educated people would agree,
But we were all so bloody tender
(And I had been reading *Dracula*)—
Then he came by, for some our supervisor,
Soon to join the Friends' Ambulance Unit.

Not famous as a friend of Chiroptera
But famously anxious and kind-hearted,
Older and wiser, a College officer.
He pulled a face as we led him to the spot.

One purblind peep, and the bat arose hangdog
And teetered off. It recognized authority.
Like Lazarus! we said: which grieved his modesty.

Alias DOA

BID they note. As if about to make an offer,
Except they seem too bored, too blasé.
Brought in we surely were, at decent speed.
Drunk, then? Faint air of disapproval. Doped?
Bad, bad, they signal. But what sort of bad?
Him they take away. Me they tell to go.
I bid good-night. We find our separate beds.

A word

But why? Why? All I ask is a word of explanation.

No comment.

A good man, yet you make him suffer like this. Or, very well, you allow it.

No comment.

Killing him inch by inch. Or allowing it. And he in full possession of his senses, at least his senses . . .

Yes.

And his dignity. Robbed even of that. The needles, catheters, bags . . .

Yes.

Such a great spirit. We need him, alive. You can hardly need him dead.

No comment.

Granted, but why this way? All we ask is a word of explanation, never mind whether we like it or not, just a word—for God's sake—

Why don't you ask him?

Oh yes, I'd forgotten. He's a Christian. But you know I can't—

Letters and life

It is remarked that the famous poet, latish in life, married
For a second and luckier time, and was happy and
  comfortable.
And wrote no more great harrowing poetry. Which goes
  without
Saying, and can be said to be distinctly remiss. For poets
Are meant to be unhappy and have many lovers (faithful
  alone
To the muse), and die relatively young. This is art's nature—
As long as you are younger than relatively young, and happy
Enough to endure the thought of others' distresses, and even
Your own future misery, tempered by years of creating great
Poetry, harrowing and famous. Later it can come to seem that
Somehow or other great poetry is rather more common than
Mere happiness.

OXFORD POETS

Fleur Adcock	Medbh McGuckian
James Berry	Jamie McKendrick
Edward Kamau Brathwaite	James Merrill
Joseph Brodsky	Peter Porter
Basil Bunting	Craig Raine
W. H. Davies	Christopher Reid
Michael Donaghy	Stephen Romer
Keith Douglas	Carole Satyamurti
D. J. Enright	Peter Scupham
Roy Fisher	Penelope Shuttle
David Gascoyne	Louis Simpson
Ivor Gurney	Anne Stevenson
David Harsent	George Szirtes
Anthony Hecht	Grete Tartler
Zbigniew Herbert	Edward Thomas
Thomas Kinsella	Charles Tomlinson
Brad Leithauser	Chris Wallace-Crabbe
Derek Mahon	Hugo Williams